AIRCRAFT

Design	Cooper — West
Editor	James McCarter
Researcher	Cecilia Weston-Baker
Consultant	Julian Moxon
Illustrator	Gerard Browne

Designed and produced by
Aladdin Books Ltd
70 Old Compton Street
London W1

First Published in
Great Britain in 1985 by
Franklin Watts
12a Golden Square
London W1

ISBN 0 86313 293 6

Printed in Belgium

MODERN TECHNOLOGY
AIRCRAFT

BILL GUNSTON

FRANKLIN WATTS/ALADDIN BOOKS

LONDON·NEW YORK·TORONTO·SYDNEY

Canadair Challenger business jets being made at Montreal

Foreword

Everyone is excited on their first flight. There's the noise of the jets on takeoff, the sight of the city streets far below and the fact that in just a few hours they'll be on the other side of the world.

Yet it was less than eighty years ago that Blériot flew the first aircraft across the English Channel. Since then, aircraft technology has made incredible advances: breakthroughs in design, electronics and other aircraft systems are bringing improvements in flight performance, economy and safety. In this book, there are examples of the latest developments in aircraft technology used in all types of aircraft, from jetliners to fighters, business jets to microlights flown for fun.

Contents

Aircraft design

① Electronics

Advanced electronics and radar systems allow pilots to fly their aircraft in all types of conditions. Contact with air traffic control is maintained by radio.

The great problem early aviators had to overcome was to produce sufficient power to give the aircraft enough forward motion to give it lift. To become airborne, an aircraft needs a force to overcome gravity – this is achieved by using the wings. An aircraft's engines drive it through the air. In addition to this, an aircraft needs systems to help the pilot navigate and communicate with people on the ground, and to ensure the greatest possible safety of passengers and crew.

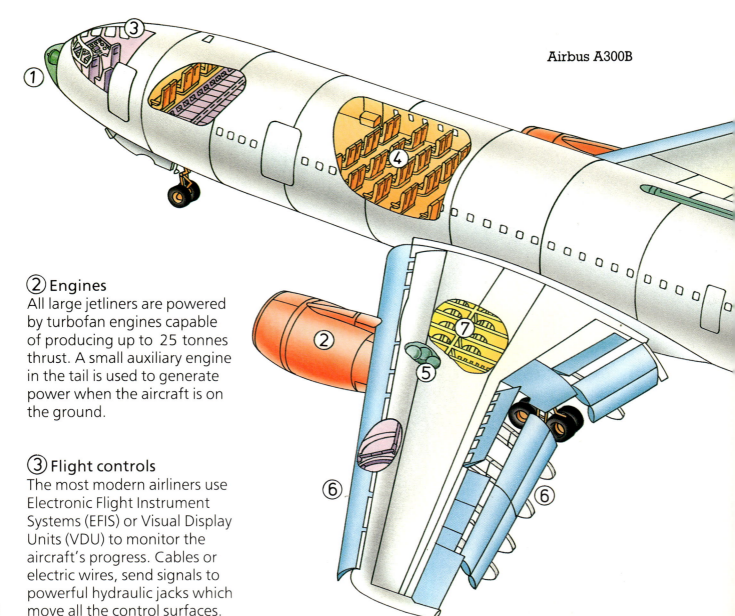

Airbus A300B

② Engines

All large jetliners are powered by turbofan engines capable of producing up to 25 tonnes thrust. A small auxiliary engine in the tail is used to generate power when the aircraft is on the ground.

③ Flight controls

The most modern airliners use Electronic Flight Instrument Systems (EFIS) or Visual Display Units (VDU) to monitor the aircraft's progress. Cables or electric wires, send signals to powerful hydraulic jacks which move all the control surfaces.

④ Safety

Before entering service every aircraft is tested for safety, for example the undercarriage is subjected to stresses greater than those experienced in flight. Inflatable escape chutes are stored inside the doors for emergency crash landings.

⑤ Fuel systems

Big airliners carry their fuel inside the wings, often from tip to tip. Automatic fire extinguishers are placed alongside the fuel tanks in case of an emergency.

The role of electronics

Basic shape, especially wing design, engines and electronics are the three major areas of modern aircraft technology. But it is the introduction of computer technology which represents *the* major advance in aircraft technology over the past ten years, both in military aircraft and airliners. Electronic systems monitor such things as fuel supply, engine performance and the conditions in the cabin. Pilots now have their course plotted and checked automatically, by "autopilot". With radar, flights can take place in all kinds of weather conditions. The diagram on this page illustrates the Airbus A300B, a European airliner used for flights of up to 4,800km (3,000 miles).

⑦ Construction

Most aircraft are built of a light metal called aluminium. Inside the smooth skin is a strong framework like a skeleton. Fibre-composites are now being used in some aircraft parts.

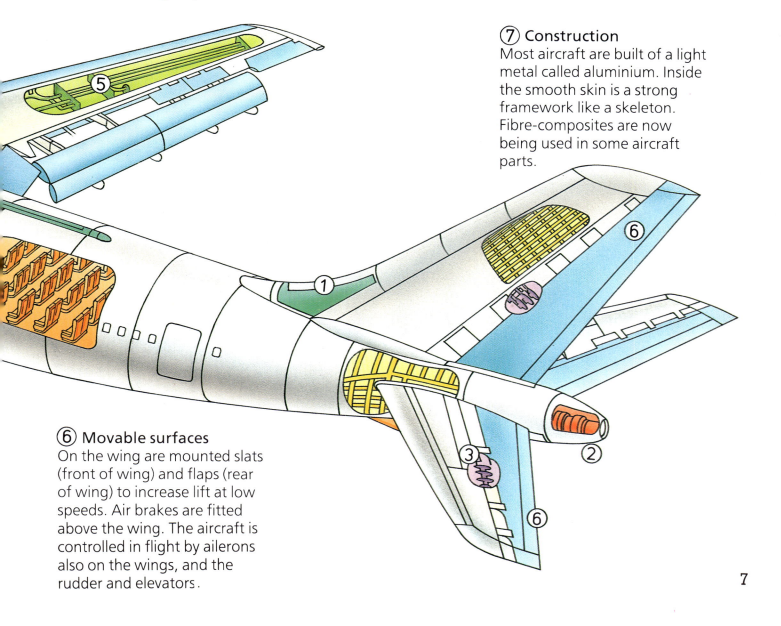

⑥ Movable surfaces

On the wing are mounted slats (front of wing) and flaps (rear of wing) to increase lift at low speeds. Air brakes are fitted above the wing. The aircraft is controlled in flight by ailerons also on the wings, and the rudder and elevators.

Principles of flight

What keeps an aircraft in the air? The answer is the speed at which it travels combined with the wing shape. The wings are shaped so that, as the aircraft moves faster and faster down the runway, the pressure is much less on the upper surface of the wing than on the underside. This difference in pressure creates a lifting force which, on takeoff, is great enough to overcome the aircraft's weight and keep it airborne.

Controlling the aircraft in the air

In flight, an aircraft's direction is controlled by altering the airflow over the wings and tail. This is done by changing the position of control surfaces – the ailerons at the rear edges of the main wing – and the elevators and rudder at the tail. These surfaces are controlled by the pilot moving the wheel, or stick (depending on the aircraft). For example, pulling the wheel back pivots the elevators up and causes the 'plane to climb. Moving it to the left or right controls the ailerons, causing the aircraft to "bank" or turn.

▽ The Boeing 747, better known as the Jumbo Jet, can carry up to 550 passengers. The four engines fixed under the wings pull the Jumbo along with a force of over 100 tonnes, but it still takes a distance of up to 3 km (almost 2 miles) for a fully loaded Jumbo to become airborne. There are over 600 Jumbos in service throughout the world.

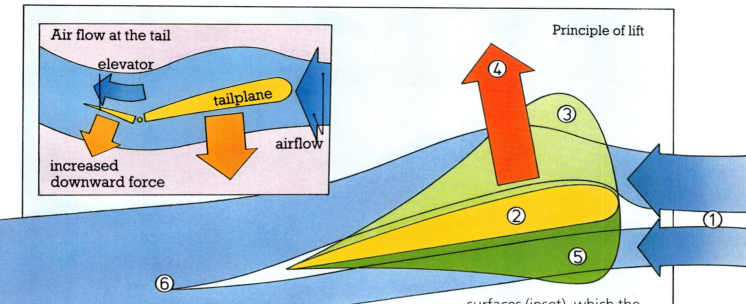

Principle of lift

Air flow at the tail

elevator

tailplane

airflow

increased downward force

△ When air (1) rushes past a wing (2) it moves faster over the top than across the bottom. This reduces the pressure on top of the wing (3), and creates suction which pulls the aircraft up into the air. The force pulling upwards is called lift. (4). Flow under the wing (5) meets the top flow again at (6). Aircraft are controlled by hinged surfaces (inset), which the pilot moves to deflect the air. The elevator is shown raised. This applies a down ward force to the tail and the nose of the aircraft is therefore pulled upwards. The change in the wing's angle also increases lift.

◁ Aircraft need three sets of controls in order to fly safely. At the ends of the wings are ailerons. When one aileron goes up, the one on the opposite wing goes down. This either makes the aircraft roll over to the left or right or holds the wings level. The tailplane carries hinged elevators (sometimes the tailplane itself is pivoted) which can make the tail pitch up or down; if the tail is pulled up the aircraft dives, and if it is pushed down the aircraft climbs. The rudder pushes the tail to left or right to create yaw.

roll-ailerons

pitch-elevators

yaw-rudder

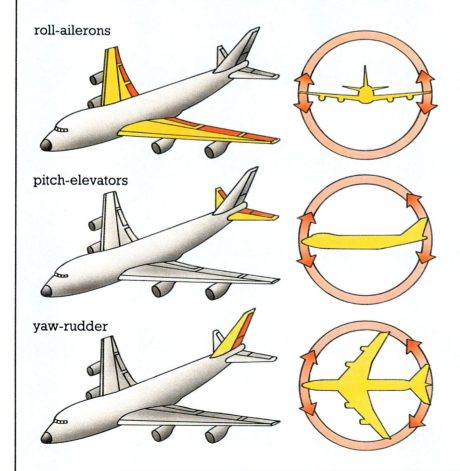

Wings

▽ Shown below are some distinctive wing shapes: Shorts 330 (1), with a conventional wing; Boeing 747 Jumbo (2), a fast jetliner with a swept wing; Mirage supersonic fighter (3), with delta wing; Concorde supersonic airliner (4); Tornado (5), with pivoted "swing wings"; Grumman X-29 (6), with the new forward swept wing.

In smaller aircraft, the wing is usually at right angles to the fuselage, but on jets, the wing is swept back, to improve performance at high speeds. Supersonic aircraft sometimes have an almost triangular wing. These aircraft are called tailless deltas. However, at lower speeds, an ordinary wing provides greater lift. A few supersonic aircraft like the Panavia Tornado have a pivoted wing that can be spread out to each side for takeoff and landing, or folded right back as a swept-wing for supersonic speeds.

Wing shapes

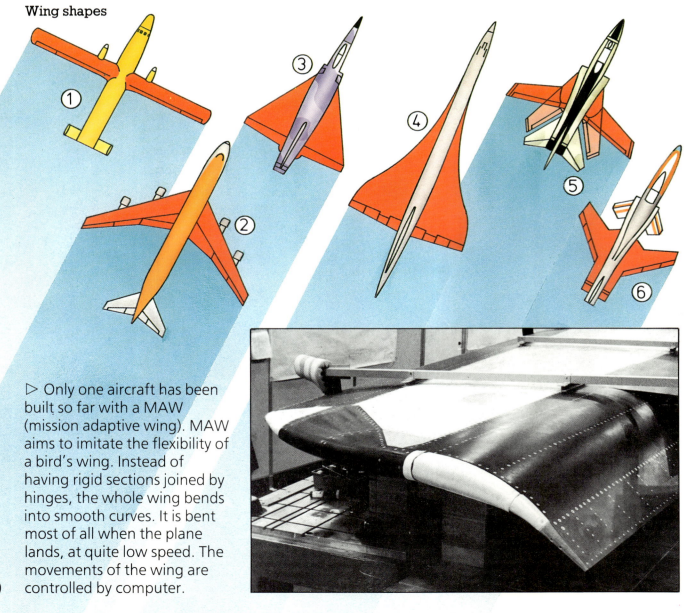

▷ Only one aircraft has been built so far with a MAW (mission adaptive wing). MAW aims to imitate the flexibility of a bird's wing. Instead of having rigid sections joined by hinges, the whole wing bends into smooth curves. It is bent most of all when the plane lands, at quite low speed. The movements of the wing are controlled by computer.

△ The X-29 has a forward swept wing. This can make a fighter very manoeuvrable, but is difficult to make, as the wings have to be strong enough to withstand enormous twisting forces.

▽ Before any new aircraft is allowed to go into production it has to be tested to prove it is safe. Here an Airbus A310 has its wings bent upwards with almost twice the force it could ever meet in flight.

Greater manoeuvrability

Fighter aircraft must be strong and manoeuvrable if they are to survive in combat, and a great deal of research is going into new wing designs. The Grumman X-29 has forward-swept wings made of the latest light but super-strong materials. The Airbus A320 range and the new generation of Boeings (757 and 767) all use wings that are designed and shaped with the help of computers, the result being a big improvement in fuel efficiency and range. They are also lighter and able to provide maximum lift over a much wider range of speeds and altitudes than previous wings.

Engine power 1

All aircraft engines operate by thrusting air backwards. Many aircraft, like business jets and trainers, are powered by turbojets. These push a jet of hot gas backwards with such force that it gives the aircraft forward speed. Turbojets burn a lot of fuel and are noisy, whereas propellers burn less fuel and are quieter. However, propeller aircraft cannot fly very fast. The turbofan combines the advantages of turbojet and propeller engines but it still allows aircraft to fly at jet engine speeds. The Boeing 767 and the Airbus A310 are each powered by two turbofans.

▷ The simplest jet engine is a turbojet. Air is drawn in at the inlet (1) and into the compressor (2). Hot compressed air enters the combustion chamber (3) The expanding gas drives the turbine (4) and forms the exhaust jet (5) which provides the thrust.

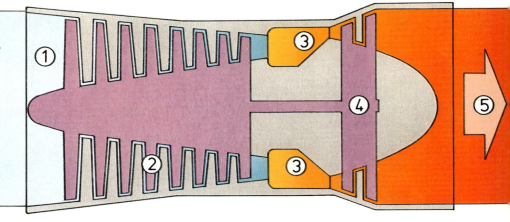

▷ Today most jets have engines of the turbofan type. Some parts, (1) to (5), are similar to a turbojet, but a turbofan also has an extra turbine (6), which drives a separate shaft to a big fan (7). This now produces most of the thrust by pumping vast quantities of air through a duct (8) around the core. Turbofans burn less fuel than turbojets, for equal thrust, and are also very much quieter.

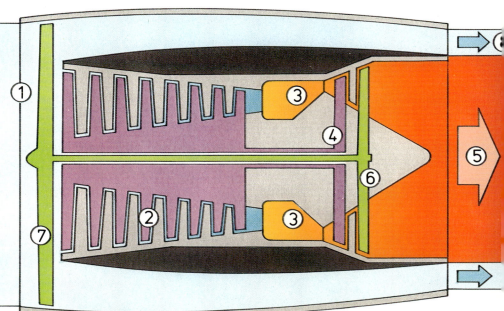

What makes a good engine?

Fuel consumption accounts for the largest single operating cost for airliners, so after safety, efficient fuel consumption is the major aim in new engine design. At present, most aircraft are fuelled with kerosene which is very expensive. For military purposes however, speed and flight performance is more important than economy. Many fighter engines have afterburners which burn extra fuel to give rapid acceleration to supersonic speeds. Afterburning, also called reheating, gives a big increase in thrust for very little extra engine weight.

▽ Tornado F.2, a new fighter for Britain's RAF, has its two turbofan engines at full power, for supersonic flight. The bright flames are in the afterburners – special jetpipes where extra fuel can be burned.

▷ Rolls-Royce produces the RB.211, a giant modern turbofan used in Jumbo Jets. At full power the fan, seen here, swallows almost a tonne of air every second, pushing the aircraft forwards at a cruising speed of 965 km/h (600 mph). This engine weighs about 4 tonnes and has a thrust at full power of over 25 tonnes.

Engine power 2

▽ Turboprops are similar to turbojet engines but they also use a propeller. Air is drawn in at the inlet (1), and into the compressor (2), to enter the combustion chamber (3). The expanding gas drives the turbine (4) and forms the exhaust jet (5). However, an extra turbine (6), is used to drive a propeller through a reduction gear (7) to generate extra thrust (8).

The first aircraft were powered by propellers, driven by a piston engine similar to that of a car. For smaller, slower aircraft, propellers are still the most economical choice. The propeller acts in a similar way as the wing, only instead of generating upward lift, it creates a forward force to move the aircraft through the air, by creating a lower air pressure in front of the blades.

Turboprops and propfans

A turboprop engine is one in which the propeller is turned by a turbojet engine – as shown in the diagram below. Turboprops have been used in commercial and transport aircraft, and their good fuel economy may make them a more popular choice in the future. In addition, a completely new type of propeller, called the propfan, is being developed.

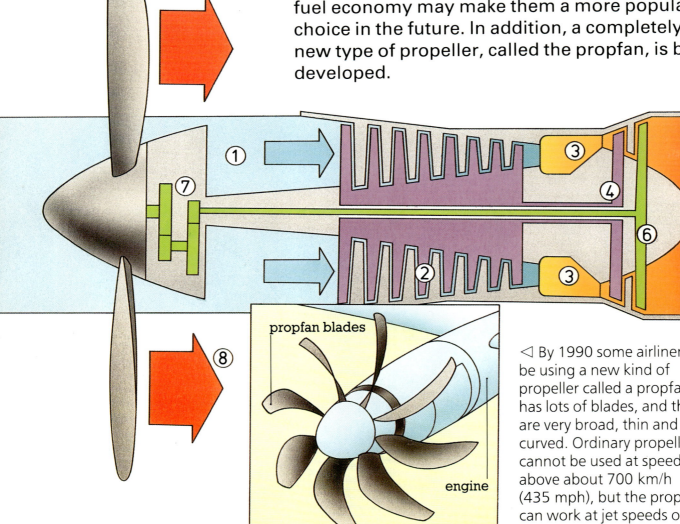

propfan blades

engine

◁ By 1990 some airliners may be using a new kind of propeller called a propfan. This has lots of blades, and these are very broad, thin and curved. Ordinary propellers cannot be used at speeds above about 700 km/h (435 mph), but the propfan can work at jet speeds of 1,000 km/h (620 mph).

△ The Beech Starship is one of many new aircraft with turboprop engines. It is unusual in having pusher propellers. These are driven by engines mounted above the wings and point backwards, so that they leave the noise behind, making the cabin quieter than in other aircraft.

Pushers versus pullers

Most propeller-driven aircraft have the propeller blades mounted at the front of the wing, so that they "pull" the aircraft. Yet in the early days of aviation, propellers were often mounted at the back of the 'plane "pushing" the 'plane through the air. Recently, designers have looked at "pushers" again. Rear-mounted propellers are found on the latest small business jets produced by several American companies.

△ This Embraer Tucano (toucan) is a trainer made in Brazil, though a new version of it is being built in Northern Ireland for the RAF. It has a turboprop at the front driving an ordinary propeller.

15

V/STOL

▽ The most successful V/STOL aircraft is the Harrier. The engine which takes in air (1), compresses it (2), and blasts it out of four nozzles (3) and (4). The four nozzles can be rotated to point downwards (5), to support the aircraft's weight, or to the rear for high-speed flight.

V/STOL stands for vertical, or short takeoff, and landing. Of course, a helicopter is a vertical takeoff aircraft, but it is much slower than conventional aircraft. The British Aerospace Harrier is a V/STOL jet that combines both advantages of vertical takeoff and speed. Its jet nozzles can be angled downward to give vertical lift, or directed backward to propel the aircraft up to the speed of sound. So far, the Harrier and the Russian Yak-38 are the only V/STOL jets to be in military service.

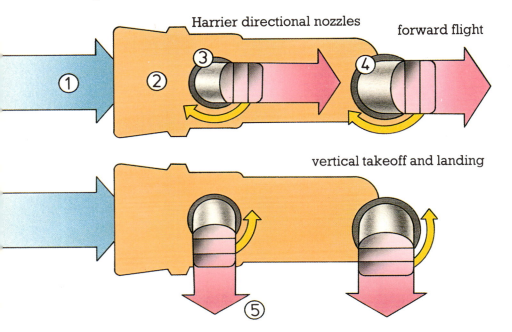

Harrier directional nozzles

forward flight

vertical takeoff and landing

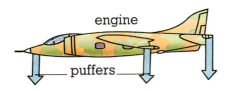

engine

puffers

△ When the Harrier is hovering, the pilot stabilises the aircraft with special high-speed air jets, "puffers", using air compressed in the engine.

Today a new version (below), the AV-8B Harrier II, is being made by British Aerospace and McDonnell Douglas USA.

△ The Bell XV-15 is a tilt-rotor research aircraft using propellers instead of jets, mounted on engines that swivel at the wingtips to provide horizontal and vertical thrust. In this picture, its propellers are tilted almost vertically so that the XV-15 can hover.

The advantages of short takeoff and landing (STOL)

For commercial aircraft, STOL means that shorter runways closer to the hearts of cities can be used. For short takeoff, extra lift is required at low speeds. This is done by increasing the airflow over the top of the wing, often by having large engines mounted high and forward. STOL aircraft also need to be quiet, so that they don't disturb local residents. The DHC-5 Buffalo, DHC-7 and the British Aerospace 146 are three successful STOL aircraft in operation today.

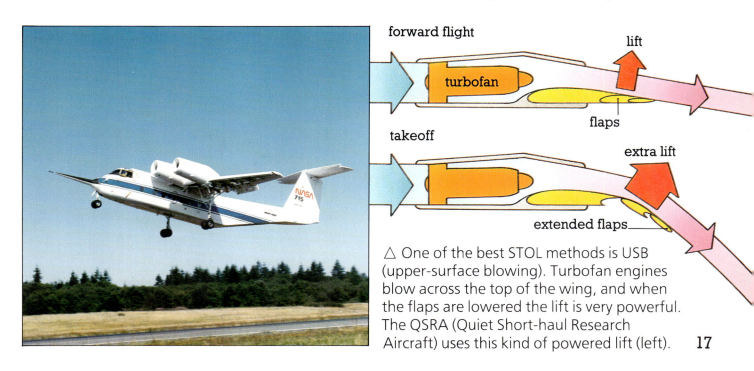

△ One of the best STOL methods is USB (upper-surface blowing). Turbofan engines blow across the top of the wing, and when the flaps are lowered the lift is very powerful. The QSRA (Quiet Short-haul Research Aircraft) uses this kind of powered lift (left). **17**

In the air

As well as actually flying the aircraft, the pilot keeps check on navigation and communicates with air traffic control. Fighter pilots also operate their weapons' systems during combat. On commercial airliners, there are usually two pilots, and sometimes a flight engineer. More and more, their job is simply to monitor the aircraft's performance using the instrument display – the on-board computer does most of the work automatically.

Pilot training

Nevertheless, a pilot undertakes extensive training. Much of this is done in sophisticated flight simulators. The simulator is an exact mock-up of a particular airliner's flight deck. Here, pilots can experience and deal with every possible emergency in complete safety.

▽ Simulators are very expensive and complicated, but they cost much less to operate than an aircraft. This one is mounted on big hydraulic struts so that it can move up and down or tilt in any direction. Projectors (1), throw a picture onto a transparent screen (2), which is reflected in a mirror (3) around the windows of the mock-up cockpit (4).

▽ This simulator has a "wrap around" external visual system which enables the instructor to recreate any type of flying condition.

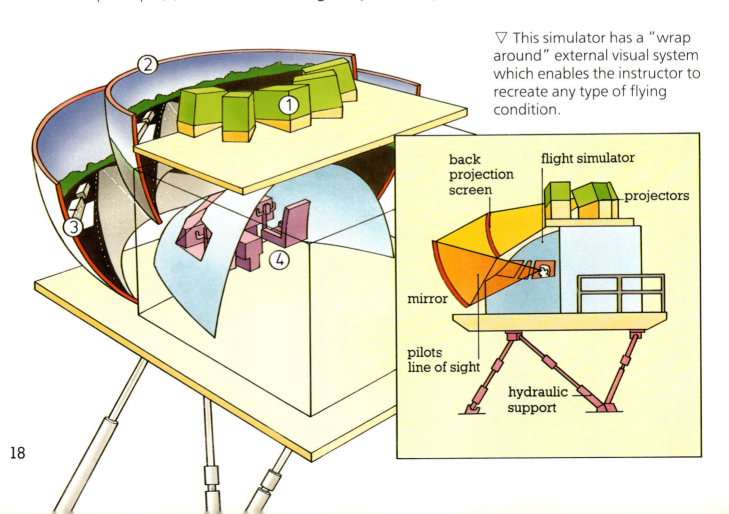

back projection screen

flight simulator

projectors

mirror

pilots line of sight

hydraulic support

▷ This airport scene was made by a computer in a VISA 2 system forming part of a French flight simulator. Because such scenes are computer-generated they can be altered in any way that the instructor or examiner wants. Day can become night, or a snow blizzard can be added.

△ The Airbus A320 has a very modern cockpit. Instead of traditional clock-like instruments, there are VDU (visual display unit) screens on which the pilots can call up any information. One screen (inset) is seen with synthetically created dials, and a moving diagram showing the flight positions of the control surfaces.

The flight deck

If anything should go wrong during a flight on a modern airliner, the pilot is informed immediately on an electronic instrument display. The display may show the aircraft's route, and the position of mountains, storm clouds or other dangers. It also gives details of the aircraft's speed and height. At the press of a button, any other information that the pilot requires can be displayed.

Safety in the air

Every day, millions of people fly safely to their destination. There are more flights now than ever before, yet airline accidents are so rare that they make front-page news. From design and construction, to day-to-day operation, aircraft builders and airline operators make every effort to eliminate every kind of possible failure.

Testing to the limit

In factory tests, an aircraft is made to suffer much greater stresses than it ever will in normal flying. Landing gear is smashed downward by huge machines, wings are vibrated and twisted in test rigs. Frozen chickens are even fired at 1,000 km/h (620 mph) into windscreens and running engines!

▷ When new aircraft are being tested they are deliberately flown badly, landed in severe crosswinds and generally tested to their limits. Here a Boeing 767 is having its tail scraped along the ground on takeoff.

▽ At the Embraer works in Brazil a new small passenger aircraft, the Brasilia, is being bent and twisted under forces much worse than anything it could encounter in flight. The metal skin is covered in special strain gauges which measure the amount of "give" in the structure.

△ Modern airliners are built so as to withstand all but the worst crash landings. The main danger has been that the fuel might catch fire. To try to prevent this, a new chemical, Avgard is being developed, which can be added to jet fuel. Here two planes have been deliberately crash-landed so that the tanks burst. The one with Avgard fuel (on the right) did not explode.

In-flight safety

All airliners carry a "black box" (in fact, it's usually painted red!) which automatically records all aspects of the aircraft's performance during flight. Should an accident occur, the black box will give invaluable information about the cause of the aircraft's failure.

The greatest danger to an aircraft on crash landing is the risk of fire. At present, researchers are trying to develop an additive to aircraft fuel which prevents it from exploding in the event of an accident. On airliners, seats and on-board equipment are made of fire-resistant materials to further reduce the hazards of fire and toxic fumes. The greatest risk in flight, however, is decompression inside the aircraft.

New materials

Most modern aircraft are built from metals, the inner structure and the outer skin consisting of a light but strong, aluminium alloy. In any large piece of metal, however, the individual metal crystals are only held together by relatively weak links. Over time, stress can cause these links to break, resulting in a tiny hair-line fracture which can threaten the structure of the whole part. Recently, "single crystal" metal parts have been developed and are already used for engine turbine blades in airliner engines. Another new technique subjects metals to high pressures and temperatures so that they can be formed into complex units without any joints that might be vulnerable to stress.

▷ Here the wing of the new Harrier II is being X-rayed to make sure that all the fibres are stuck together perfectly.

▽ The French Dassault-Breguet Rafale is a new fighter for use in the 1990s. The colours show some of the special materials it is made of. The carbon-fibre parts are made with millions of fibres, set in a tough resin and arranged, like plywood, in different directions.

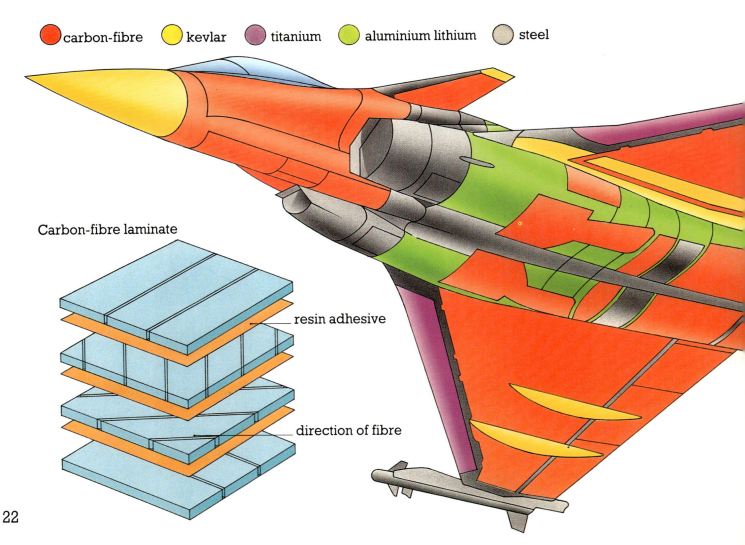

● carbon-fibre　　○ kevlar　　● titanium　　● aluminium lithium　　○ steel

Carbon-fibre laminate

resin adhesive

direction of fibre

Getting away from metals

Over the last twenty years, new, non-metallic materials have been developed. These are called "composites". One of the most widely used is carbon-fibre, stronger than steel and much lighter. Another new composite is Kevlar. A single hair-thick strand of Kevlar can support the weight of an adult. Kevlar is being used in the latest military aircraft, such as the prototype French Rafale fighter shown opposite.

▷ New composite structures are "baked" in an "autoclave" like the one in the photograph, to improve their strength. An autoclave is a mixture of a giant oven and a press.

Warplanes

An advanced, modern airforce uses a wide range of different types of aircraft, each designed to perform a specific role in time of war. There are transport aircraft for carrying men and equipment to a battle zone, spy planes that are airborne for long periods seeking to detect enemy positions and troop movements, and a host of smaller 'planes for routine activities. But those at the front line of combat, the fighters and the bombers, are among the most sophisticated machines ever built.

A lethal combination

Today's fighter is packed with advanced electronic systems, loaded with the latest missiles and engages the enemy at speeds of up to 2,000 km/h (1,240 mph). The latest development is a head-up-display for the pilot — all the information he needs is projected on the cockpit window in front of him by the computer.

▽ A HUD (head-up display) tells a pilot vital information without making him look down into the cockpit or refocus his eyes. The HUD of the Harrier II (inset) is projected on the glass plate at the top between the dark lines.

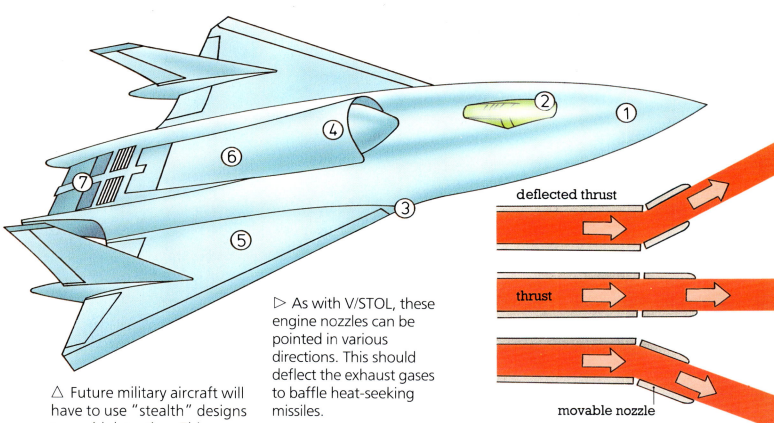

▷ As with V/STOL, these engine nozzles can be pointed in various directions. This should deflect the exhaust gases to baffle heat-seeking missiles.

deflected thrust

thrust

movable nozzle

△ Future military aircraft will have to use "stealth" designs to avoid detection. This illustration shows a fighter of the future with many new features: flat sharp-edged nose (1), small cockpit with non-reflective canopy (2), flat underside (3), engine on top, with radar-absorbent paint on skin (4), thin delta wings with radar-absorbent paint (5), smooth flat top (6), special engine nozzles (7).

▷ Apart from the Soviet Union, the US is the only country developing large-size bombers. The B-1B entered service with the US Air Force in 1985. Though about the same size as the B-52, which it is replacing, it has an appearance on enemy radars only one-hundredth as large, because it incorporates partial "stealth" technology.

Hiding from the enemy

New "stealth" technology has been developed for the latest generation of bombers, so that they can pass enemy electronic detection systems unnoticed. Stealth fighters and bombers are carefully shaped so they do not easily reflect radar signals and have engines designed to emit a minimum of infra-red heat.

Aircraft contrasts

The great majority of the 20,000 or so aircraft built every year go to neither the commercial airlines nor the armed forces. These are smaller aircraft, used for business, farming or fun.

Smaller aircraft at work

Business or executive jets are usually powered by two or three small turbofan engines and have a maximum range of about 7,000 km (4,200 miles). For shorter ranges, small turboprops such as the Beech King Air are popular. Agricultural aircraft are designed for the rough and ready working life they lead. Cockpit design allows the pilot to see the land beneath him and side panels can be removed so that the structure can be hosed down to remove chemicals used in crop spraying.

▷ Today, there are still a few large aircraft powered by piston engines. One is the Canadian CL-215, which can land on an airfield or on water. One of its jobs is to scoop up water and drop it on forest fires. By contrast, the little single-seat autogyro (far right) is used just for fun.

▽ Microlights are powered by very small piston engines driving a propeller. This one has two tiny engines just in front of the wing. Many microlights have a transparent plastic skin.

△ Airbus components are made in factories all over Europe, the body in Germany, wings in Britain and other parts in Belgium and Spain. The parts are brought to Toulouse in southern France in giant cargo aircraft called Super Guppies.

Aircraft for sport and leisure

People have flown for fun and raced against one another since aviation began. Many people begin their flying by joining gliding clubs before moving on to powered aircraft. Today home-constructed microlight aircraft are no more expensive than a second-hand car. A microlight is basically a flexible wing, powered by a petrol-driven engine. They have a range of about 100 km (62 miles), covered in about an hour's flying time. In addition, many enthusiasts are still trying to build ultra-light machines that can be flown under man-power alone.

27

Towards 2000

Many of the new ideas for aircraft technology already discussed in this book are still at the experimental stage. They will undergo years of testing and modification before they are put into everyday use. Some may never go into production. For instance, there are many ideas for supersonic jetliners, bigger and faster than Concorde. But with the present high cost of fuel, the airlines are likely to ignore them for the next twenty years.

Conventional aircraft fuel – kerosene, made from oil – is likely to remain expensive, and one day our supplies of oil will run out. So a great deal of research is taking place to find fuels to replace kerosene. Liquid hydrogen looks promising, but takes up a great deal more space. Hydrogen-fuelled aircraft would have to be fatter, or carry large fuel pods beneath their wings.

▽ Ames-Dryden AD-I is the NASA-funded research aircraft built to investigate "slewed wing" technology. Unlike swing wing aircraft, the slewed wing is pivoted in the middle: when one half of the wing goes back the other goes forward. A one piece wing gives the benefits of less weight; it also means the wing can be arranged conventionally on takeoff and swept back at supersonic speeds.

△ One of Lockheed-California's possible designs for the 1990s. This sleek aircraft would cruise at 5,470 km/h (3,350 mph) at a maximum height of about 30,480 m (100,00 feet). The aircraft would be made primarily of titanium with its outer edges constructed of heat-resistant stainless steel. At top speed, the leading edges of the airframe would glow red.

New designs

Some think that giant airships may be revived to carry heavy cargoes. Another idea is the Spanloader, basically a giant wing with freight containers strapped inside. Perhaps, with computers becoming more powerful each year, the day of the unmanned robot aircraft is not too far away. Such aircraft would simply be monitored from takeoff to landing by air traffic control staff on the ground.

▷ The ATP (Advanced Technology Turboprop) is a new British airliner which is expected to make its first flight in August 1986 and be ready for delivery in September 1987. Seating 64 people, it is driven by two six-bladed propellers.

Datechart

December 17 1903

The first man-carrying powered flight is made by Wilbur and Orville Wright at Kitty Hawk, USA. The flight lasted twelve seconds and covered 40 m (120 ft).

July 25 1909

Louis Blériot flies in his monoplane from France to England.

October 22 1911

Capt Piazza makes a reconnaissance flight over Turkish troops in North Africa. A week later another Italian pilot drops bombs. These are the first aircraft flights made in war.

February 5 1919

Start of the first sustained daily passenger airline service, operating between Berlin and Weimar.

June 14/15 1919

Capt J. Alcock and Lt A. Whitten Brown make the first non-stop crossing of the Atlantic, from Newfoundland to Ireland. The total flying time was 16 hours 27 minutes.

May 20/21 1927

Charles Lindbergh flies non-stop New York to Paris, in his monoplane *Spirit of St. Louis*.

July 12 1944

The Gloster Meteor MkI becomes the first jet aircraft to enter operational service.

October 14 1947

Capt Charles "Chuck" Yeager flies the rocket-propelled Bell X-1 just beyond the speed of sound in the world's first supersonic flight.

July 16 1948

First flight of the Vickers Viscount, the world's first turboprop airliner.

July 27 1949

First flight of the de Havilland Comet, the first jet airliner.

October 3 1967

North American X-15A-2 flown by William Knight reaches 7,297 km/h (4,534 mph), the highest speed ever achieved by an aircraft.

January 21 1976

The world's first supersonic passenger services are opened by simultaneous departures of British and French Concordes for Bahrain and Rio de Janeiro.

June 1984

Introduction of non-stop services from Hong Kong to London, 12,000 km (7,500 miles) with Jumbos.

1985

Concorde sets two new world records for passenger airliners – February 13, London to Sydney in 17 hours 3 minutes, and March 28, London to Cape Town in 8 hours 8 minutes.

Glossary

Ailerons Hinged surfaces on the rear edge of the wing, used to roll the aircraft or hold the wings level.

Delta Triangular shaped wing. A delta wing flies point first, and often has elevators on its trailing edge.

Fuselage The main body of an aircraft.

Head-up display A special cockpit device directly in front of the pilot which projects many kinds of information on a transparent glass screen. The pilot does not have to refocus his eyes to read the information.

Propfan A multi-bladed propeller with curved blades like scimitars. It promises the economy of the propeller with the speed of the jet.

Rudder The hinged vertical control surface attached to the fin.

Stealth New techniques for making aircraft as invisible and undetectable as possible.

Supersonic Faster than the speed of sound in the surrounding air. The speed of sound (called Mach 1.0) varies from about 1,225 km/h (760 mph) in warm air at sea level to about 1,060 km/h (661 mph) in very cold air at great heights.

Tailplane The horizontal part of the tail. It may be fixed or used as a movable control surface.

Turbofan A turbojet engine with a fan added to give extra thrust. Unlike a turbojet, much of the thrust is caused by the expelling of air under pressure.

Wingspan The distance from wingtip to wingtip.

Index

Acknowledgements

The publishers would like to thank the following organisations who have helped in the preparation of this book:
Airbus Industrie, Alitalia, Benson Aircraft Corp., Boeing, British Aerospace (BAe) — Bracknell, Hatfield, Kingston and Manchester Divisions, British Aerospace Dynamics Group, British Caledonian, Canadair, Civil Aircraft Authority, Dessault-Breguet Aviation, Dowty Group Services, Embraer, Fairchild Republic Corp., Gates Learjet Corp., General Electric (US), Goodyear, Gulfstream Aerospace, ICI Paints Division, Interavia, Japan Air Lines, Japanese Tourist Organisation, Lufthansa, Marconi, McDonnell Douglas, Pilatus Aircraft, Public Relations Bureau on behalf of Lockheed International, Rediffusion, Rolls-Royce, Sabena, Smiths Industries, Thomson CSF, United Technologies Pratt and Witney, Qantas and thanks to Dorchester Typesetting.

Photographic Credits:

Cover: Airbus Industrie; *title page*: Quadrant Picture Library (QPL); *contents page*: Canadair; page 8, Qantas; page 10, Boeing; page 11, Michael Taylor, Airbus; page 13, QPL, Rolls-Royce; page 15, QPL, Embraer; page 16, BAe; page 17, Michael Taylor, Michael Taylor; page 19, Thomson CSF, Airbus; page 20, Embraer; page 21, Boeing, ICI; page 23, McDonnell Douglas, Embraer; page 24, Art Directors, Smiths Industries; page 25, Michael Taylor; page 26, Michael Taylor; page 27, Canadair, Benson, Airbus; page 28, Michael Taylor; page 29, Lockheed, BAe.

PRINTED IN BELGIUM BY

proost
INTERNATIONAL BOOK PRODUCTION